The Long-Standing Scientific Essays

Miguel A. Sanchez-Rey

Table of Contents

Immeasurable Ecology

3-9

Advances in the Collective Interface

11-15

An experimental work of literature

17-24

PHPR

26-27

The Scientific Age: Protocol

29

PHPR Protocol

31

Ad [superstrings] : Incalculable particles

33

Immeasurable Ecology

[Author: Miguel A. Sanchez-Rey]

The Search for Extraterrestrial Intelligence [SETI] has so far been an abysmal failure. SETI has produce no tangible or credible indirect observational data that intelligent life exists beyond ones' solar habitat. No radio signals consistent with an industrial and highly technological civilization has been detected and even with the Kepler satellite, in which evidence has been produce that Earth-like planets are more prevalent than thought, it could take centuries for a signal from those Earth-like planets to arrive at planet Earth. But even then evidence of intelligent life in those Earth-like planets, in form of radio and digital signals, should have been detected by now. So far evidence of intelligent radio signals has remained a fairly empty vacuum. Let alone digital signals.

Wild speculation in the Scientific Age has hinted that intelligent life lies more prominently at the spiral edge of the Milky Way galaxy. A part of the Milky Way that is not subject to the turbulent and hot space weather that is seen in the center of the Milky Way in which lies a massive black hole that collects much of the galaxies matter that ultimately forms its spiral shape.

Yet it's more eloquent and efficient to see the spiral edge, in conjunction with what can be gathered from the Kepler data, as containing habitable zones suitable for human settlement and outer habitable zones that are off-limits. Those habitable zones will, in its most theoretical form, contain minerals and energy resources that can be use for the development of space-habitats and biospheres by utilizing the terraformic process. Habitable zones that can only be claim upon completion of The Second Task of PHPR [The Physicalist Program] which will, at present, endeavor to realize the internationalist model.

At that point Earth will become a vestige of a crude terraformic machine. A planetary biosphere in which humanity looks helplessly at the stars at its isolation and vulnerability to self-extinction and cosmic catastrophe. That said humanity in the near-distant future cannot pin its hopes on the biosphere, whether planetary or the space-habitat, but must consider that the way forward is to increase survivability through co-existence, and technological advances that lead to more efficient and eloquent biospheres in which utilizing the transformative process for achieving habitable planetary biospheres is the last resort.

Social ecology dictates the intrinsic relationship between one's ecosystem and social organization. With the Omega-Kardeshev scale it becomes easier to interrelate sociological, psychological and technological development. But to head further into a Class 3 civilization one arrives even further into an incalculable stage. An incalculable stage where all notions of scientific measurement breaks down. If that's the case, then it's an immeasurable ecology.

An immeasurable ecology that is unknown to present-day science. In which SETI's radio telescopes and all other astronomical instruments i.e., optical telescopes and satellites, will not be able detect any direct evidence of a Class 3 civilization. A Class 3 civilization that resides as an incalculable technological society that is above all qualitative forms of present-day measurement, including observation. One primary reason is that, as astronomers have speculated in a superficial sense, is that this Class 3 civilization could reside within the vicinity of the galactic center.

Harnessing energy that is being emitted as Hawking-radiation and also exploiting red giants -- by utilizing Incalculability, that allows for large-scale use of sustainable nuclear fusion.

But indirect hints of such a civilization could be found by looking, by the mid-21st century, for twinkles that look very much like twinkles that amateur astronomers observe today in the night sky. Twinkles that move around like traffic lights seen above ground during plane flight on planet Earth.

Those twinkles, at the vicinity of the galactic center, may be indirect and yet tangible evidence of a vibrant intergalactic civilization. Keep in mind that what may be observed as twinkles will not reveal the ecosystem of an incalculable civilization. Primarily because the current laws of physics breakdown as one approaches closer to the event horizon of a black-hole and with Ad [superstrings] the current laws of physics, as well, are bound to change with increasing progress in Advance Physics. Which implies that it could take half-millennium before any clean observation is made indicative of a Class 3 civilization near the galactic center.

And for that matter the social ecology of the Scientific Age is too early to tell. One can only speculate at this point but as such it becomes clearer with the Omega-Kardeshev scale that the capacity to disassociate from the biological processes of evolution in order to withstand the stressors of an intergalactic civilization, of said nature, means the futility of both the Omega-point and the transcendence of Darwinian evolution. Instead the processes of biological life are subject to change as one arrives closer to Class 3 -- most likely as a civilization indicative of 2AI, that eventually becomes an intergalactic civilization made up of star-light. An intergalactic civilization which is currently immeasurable to a Class 0.7 civilization at the beginning of the Scientific Age that within 100 years is to arrive at Class 1.

A Class 3 civilization chooses to be immeasurable to a Class [0, 1, 2] civilization since a Class [0, 1, 2] is not ready to comprehend its incalculable stage. But gathering more coherent evidence of a Class 3 civilization at the incalculable stage should become easier, as the next few centuries transpire, that leads to further progress in Advance Physics and astronomy in the Scientific Age.

Advances in the Collective Interface

The Physicalist Program

[Author: Miguel A. Sanchez-Rey]

The collective interface is the rudimentary building block of advance consciousness. In which one's self-conscious awareness can log in and log out of the collective subconscious. A byproduct of early advance artificial intelligence and an essential element in the self-actualization of anticipatory learning machines. Anticipatory learning machines that can adapt to any obstacle and environment that is presented with that will lead to an advance technological stage. That said, advance consciousness is dissociated from the transcendental consciousness of the singularity in that transcendental consciousness eventually results in a mutilated awareness.

As the collective consciousness is a futile attempt to achieve self-actualization of transcendental Darwinism by accomplishing a consciousness that is the sum of all individual conscious-states. That is a consciousness that is indicative of artificial superintelligence. Yet superintelligence can be disadvantageous to natural selection when superintelligence supersedes its own self-intelligence.

Eventually leading to a breakdown in machine learning and adaptation. That said the lived body experience is of a higher qualitative order. A higher qualitative order that includes the interaction between the nervous system and the conscious state of self-awareness. That is by utilizing anticipatory learning methods, that can achieve maximization of adaptation, machine intelligence can bypass any higher-complex obstacle that is confronted with. Not only by applying its own superintelligence but also by utilizing the lived-body experience to accomplish a giving task.

Task meaning that an anticipatory learning machine must not only anticipate a solution to such a task but must also act in anticipation of its resolution. A resolution that perfects itself and can be perfected in such a way that it includes all possible engineering obstacles and tasks that solves all possible obstacles and tasks that leads to a breakthrough in both higher-order machine learning and engineering. A machine must not only think and adapt but must also adapt and think in such a way that it achieves its task. That task is what one calls the nature of the lived-body experience.

For the brain is not only a thinking-organ but also a task-master that through epigenetic natural selection has evolved into a fully form organism that can walk, talk and play in such a way that it can self-actualize. Integrating machine learning with human intelligence that leads to progress in cybernetics that eventually evolves into a collective interface that accomplishes an advance consciousness.

An advance consciousness that leads to an advance technological stage in human biological development. Where the processes of biological evolution are no longer of gradual molecular biology but instead epigenetic and explosive in its power to anticipate the costs of superintelligence. To consider itself a superintelligence that is of a higher order in its design and structure.

Advances in the collective interface will achieve a self-consciousness that is aware of both the collective subconscious and advance consciousness. That can shift between both the collective subconscious and the collective consciousness. That can merge both the collective subconscious and collective consciousness that is a collective interface of an advance intelligence that supersedes machine superintelligence and can be said to be far more efficient and eloquent than any machine learning methodology that includes its overemphasis on intelligence, and its requirement that it be both a memory sorting and learning machine.

That is in many respects the ideal of advance artificial superintelligence but in reality is an advance artificial intelligence that is an anticipatory adaptation cybernetic machine that does not overestimate it own self-intelligence but overcomes its own overestimation of self-intelligence towards a collective interface that accomplishes an advance consciousness that ushers an advance technological stage.

An experimental work of literature

Miguel A. Sanchez-Rey

Isaac's Laws is an experiment that ascertains and establishes the breakdown of the scientific process through a critical examination of Post-Modernity. Using literary techniques to delve into the collective psyche of the modern world. To resolve long-standing issues by applying a unique experimental apparatus meant to tap into the collective unconscious of Post-Modernism. Culminating with a final conclusion: radicalism and extremism of Post-Modernity led to a breakdown in the scientific process.

Radicalizing the general public into crying havoc when the scientific process did not conform with their conclusions of scientific theory. Using the media, establishment papers and the world-wide web to facilitate cry havoc to soot their fraudulent interests for fame and fortune. Tampering with the blind-nature of the scientific process.

That is flaw-decision making led to cry havoc, and that in which cry havoc has unleash mass opposition to both the state and the scientific process. Blurring the lines between expert decision-making in the scientific process and public participation in the sciences. In which wide-spread use of the world-wide web led to a cult-phenomenon of flaw-decision making at the dawn of the Windows XP revolution.

Flaw-decision making that cause a world-wide deterioration in the market economy and flaw-decision making that spark international conflicts. Leading to a global economic crisis, and instigating the rise of organized and/or state terrorism and ultra-nationalism. Which, having been radicalize by the radicalism and extremism of the political process, quickly radicalize the scientific process -- cause by the pillage and plunder of the masses, resulting in a serious breakdown at CERN, ITER, and eventually at LIGOS, in which the planet reach the Scientific Age in the form of an uncontrollable scientific machine.

Radicalism and extremism that is due to the mass opposition against classical liberal social and political thought. Classical and liberal social and political thought that is grounded on both the Age of the Enlightenment and the scientific method.

That is at a certain point the collective unconscious went into a global meltdown, in the form of a schizophrenic breakdown, that slipped the planet into collective psychosis that morph the neo-Fascist and social justice movement into a neo-Fascist alliance. Unleashing quiet mayhem upon the planetary biosphere -- in which, flaw decision-making gave way to a world-wide pathological mass-movement and eventually the apprehension of the scientific prize system.

A prize system that sought to preserve Western European liberal democracy when social democracy has since then collapse, with the 2016 Brexit, that led to the permanent decline of the religious state after the 2016 American Presidential Elections.

After which the Earth's biosphere reach its breaking-point signaling the decline of the planetary ecosystem, and the disintegration and fall of the norms of Post-Modernist social, ethical and political theory. Plunging the scientific process into complete disarray.

Leading to the horrific outcome of an insane planet, in collective psychosis, that destroys itself with *terraforming* by misusing ITER to bring about a short-live dynasty class and environmental boom that eventually leads to a global drought -- or a dead planet.

The horrific outcome led to the founding of PHPR [The Physicalist Program] as a resolution to a foreseeable catastrophic scenario in the Scientific Age in the form of a task. In which The First Task is a 100 Year Task that aims to resolve mineral depletion by completing and utilizing The Grand Unification Scheme.

In conjunction all the national laboratories are to pull their scientific repositories, and impose a lock-down on all scientific research pertinent to the Scientific Age and metaspace.

PHPR is to be fully implemented when ITER goes on the manufacturing line whereby the planetary biosphere will be put into a serene state for a 40-year window of opportunity of gradual environmental recovery. After which when ITER goes off-line the major powers will make the decision to cease hostilities slipping the planet into a serene and quiet state. Whereby 60 years remain to reach full and careful completion of The First Task. Achieving complete environmental recovery at the beginning of the 22nd century.

The Grandmaster is to complete a task and set the next task. PHPR is to be dismantled when the last task is completed. By then the Scientific Age has come to an end.

The unconscious makes full recovery from collective psychosis. Ushering Anarcho-syndicalism in which scientific participation is reopen to the general public at the beginning of the Advance Age -- such that, after approximately 100 years' one reaches a Class 2 civilization.

An experimental work of literature not only aims to experiment by using literature but also to make significant advances in the scientific process. The prime conclusion of *Isaac's Laws* is that the radicalization of the sciences is the consequence of the radicalism and extremism of Post-Modernity cause by the cry havoc of flaw and risky-decision making of both ideological and financial interests with fraudulent motives for fame and fortune.

Scientific radicalization entailed that the planet be pacified by shutting-off the general public from the Scientific Age until full disclosure is appropriate to the scientific process. And that PHPR's tasks not be fully disclosed until completion is in the near horizon.

Using a world-wide command economy to keep the general population from revolting any further against the Scientific Age and by also using costless education to motivate the scientific process while increasing academic standards to put an end to a rising world-wide pathological mass-movement.

The Scientific Age is bound to happen. And so the interplay of multiple strategies, between the psycho-analytic and behavioral, led to it's finality: wild-strength defines the limitations of the foreseeable in the experimental process and its logical intuition.

PHPR

(Procedural Protocol)

By

Miguel A. Sanchez-Rey

Government is a resolution to the state-of-nature [1]. The Physicalist Program [PHPR] is design as a resolution to a foreseeable catastrophic scenario in the Scientific Age in the form of a task. The Grandmaster is to complete a task and set the next task.

The First Task of PHPR is a 100 Year Task [2].

When the International Thermonuclear Experimental Reactor [ITER] goes online the planetary system will begin to gradually recover. A 40-year period of sustainable economic growth and environmental recovery sets a window of opportunity to complete 60 percent of The First Task of PHPR. After which there will be a global decline when ITER is off the manufacturing line.

PHPR's task is not to be fully disclosed until completion is in the near horizon.

PHPR is to be dismantled when the last task is completed. By then the Scientific Age has come to an end…

References

[1] Hobbes, Thomas. Leviathan. Oxford University Press: 1996.

[2] Sanchez-Rey, Miguel A. The Physicalist Program. Createspace: 2015.

The Scientific Age: Protocol

By

Miguel A. Sanchez-Rey

The Scientific Age is an uncontrollable scientific machine.

Gaining control of the scientific machine requires that at a certain point the general public be periodically shut-out from the Scientific Age so as not to cause any further harm to the scientific process.

All national laboratories research repositories are to be withheld from the general public. All scientific advances relevant to PHPR, and the Scientific Age, is to follow until full-disclosure becomes relevant to the scientific process.

Knowledge of the Advance Age is to be ascertain and establish by The Grandmaster in anticipation that the last task will be completed.

The Scientific Age leads to the Advance Age when the scientific planetary super-state is dismantled. At that point the scientific machine becomes controllable.

The Advance Age is an age of wild strength…

PHPR Protocol

Government is a resolution to the state-of-nature. The Physicalist Program [PHPR] is design as a resolution to a foreseeable catastrophic scenario in the Scientific Age in the form of a task. The Grandmaster is to complete a task and set the next task.

The First Task of PHPR is a 100 Year Task.

PHPR top-scientists are task to complete The First Task.

PHPR top-scientists are selected into PHPR.

Selection of top-scientists into PHPR is a stringent task.

PHPR top-scientists exemplify combat leadership and excellence in scientific scholarship.

Identity of PHPR top-scientists are to remain classified until their task in The First Task is partially declassified.

By then top-scientists are to retire from PHPR.

Acknowledging PHPR top-scientist as The Master of Space-Time.

The Grandmaster will continue on…

Ad [superstrings] : Incalculable particles

[Author: Miguel A. Sanchez-Rey]

Ad [superstrings] are incalculable particles. But by incalculable particles one means particles that are unlike stringy particles and/or one-dimensional objects that vibrate on a 11-dimensional supermanifold. They are not of qualitative D-branes which are D1-D6 membranes that allow for the propagation of open strings. Instead they can be seen as one-dimensional variant [of stringy]'s of prime that exist in metaspace. That is Ad [superstrings] are one-dimensional variations that exist on a 11-dimesional Calabi-Yau supermanifold.

That can replicate, twist, twirl and turn in such a way that closed strings make up bosons, and open strings constitute fermions and the graviton. In which, splitting an Ad [superstring] is indicative of metaspace. That is SUSY-like physics resides in metaspace. Ad [superstrings] are variations of one-dimensional superstringy-particle that exist in metaspace.

www.ingramcontent.com/pod-product-compliance
Lightning Source LLC
Chambersburg PA
CBHW062208220526
45470CB00009B/2971